MYTHICAL CREATURES
KRAKEN

Prospect Heights Public Library
12 N Elm Street
Prospect Heights, IL 60070
www.phpl.info

BY THOMAS KINGSLEY TROUPE

BELLWETHER MEDIA • MINNEAPOLIS, MN

TORQUE™

Torque brims with excitement perfect for thrill-seekers of all kinds. Discover daring survival skills, explore uncharted worlds, and marvel at mighty engines and extreme sports. In *Torque* books, anything can happen. Are you ready?

This edition first published in 2021 by Bellwether Media, Inc.

No part of this publication may be reproduced in whole or in part without written permission of the publisher.
For information regarding permission, write to Bellwether Media, Inc., Attention: Permissions Department,
6012 Blue Circle Drive, Minnetonka, MN 55343.

Library of Congress Cataloging-in-Publication Data

Names: Troupe, Thomas Kingsley, author.
Title: Kraken / by Thomas Kingsley Troupe.
Description: Minneapolis, MN : Bellwether Media, 2021. | Series: Torque : mythical creatures | Includes bibliographical references and index. | Audience: Ages 7-12 | Audience: Grades 4-6 | Summary: "Amazing images accompany engaging information about kraken. The combination of high-interest subject matter and light text is intended for students in grades 3 through 7"– Provided by publisher.
Identifiers: LCCN 2020046869 (print) | LCCN 2020046870 (ebook) | ISBN 9781644874653 (library binding) | ISBN 9781648341427 (ebook)
Subjects: LCSH: Kraken–Juvenile literature.
Classification: LCC QL89.2.K73 T76 2021 (print) | LCC QL89.2.K73 (ebook) | DDC 594/.58–dc23
LC record available at https://lccn.loc.gov/2020046869
LC ebook record available at https://lccn.loc.gov/2020046870

Text copyright © 2021 by Bellwether Media, Inc. TORQUE and associated logos are trademarks and/or registered trademarks of Bellwether Media, Inc.

Editor: Rebecca Sabelko   Designer: Josh Brink

Printed in the United States of America, North Mankato, MN.

TABLE OF CONTENTS

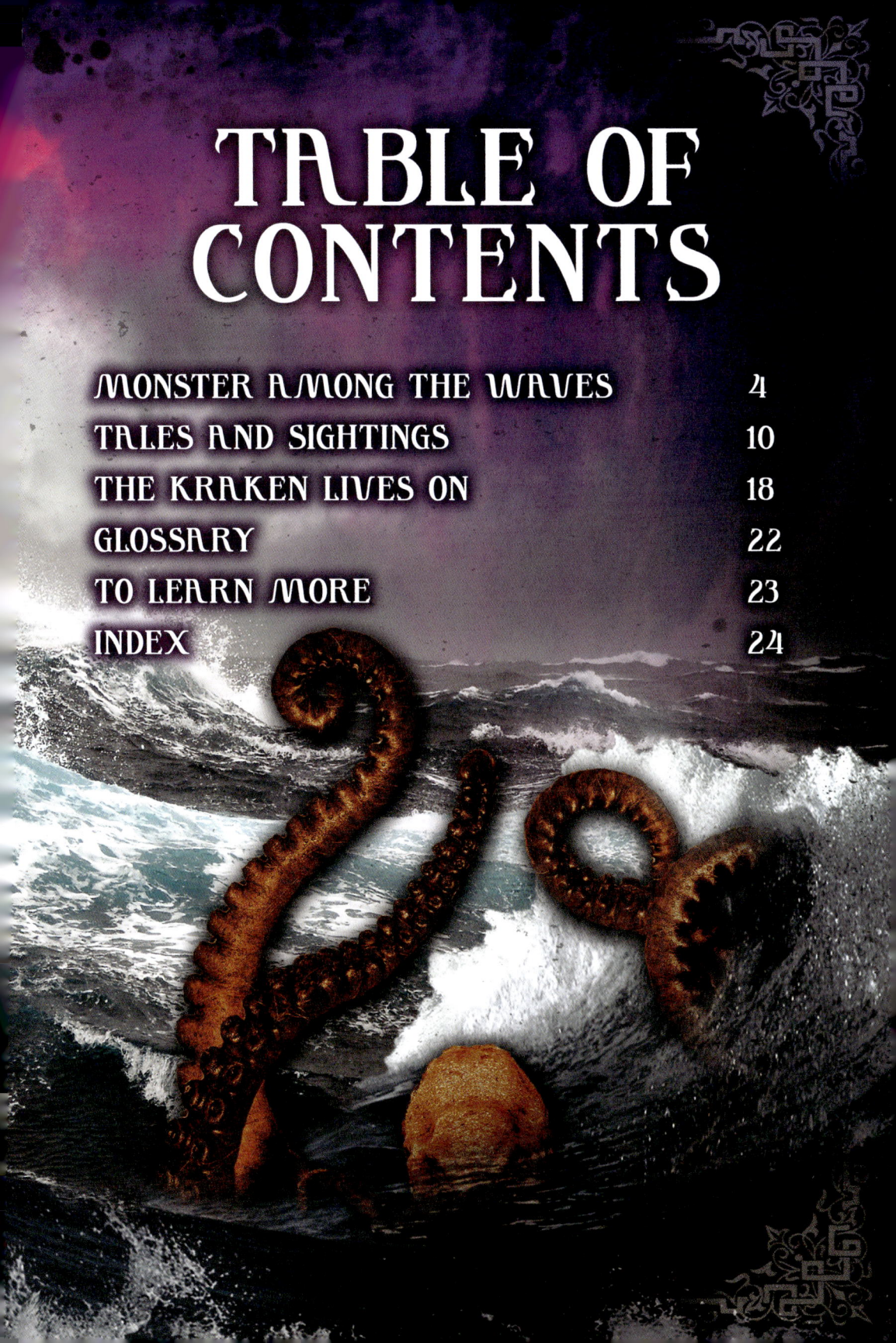

MONSTER AMONG THE WAVES	4
TALES AND SIGHTINGS	10
THE KRAKEN LIVES ON	18
GLOSSARY	22
TO LEARN MORE	23
INDEX	24

MONSTER AMONG THE WAVES

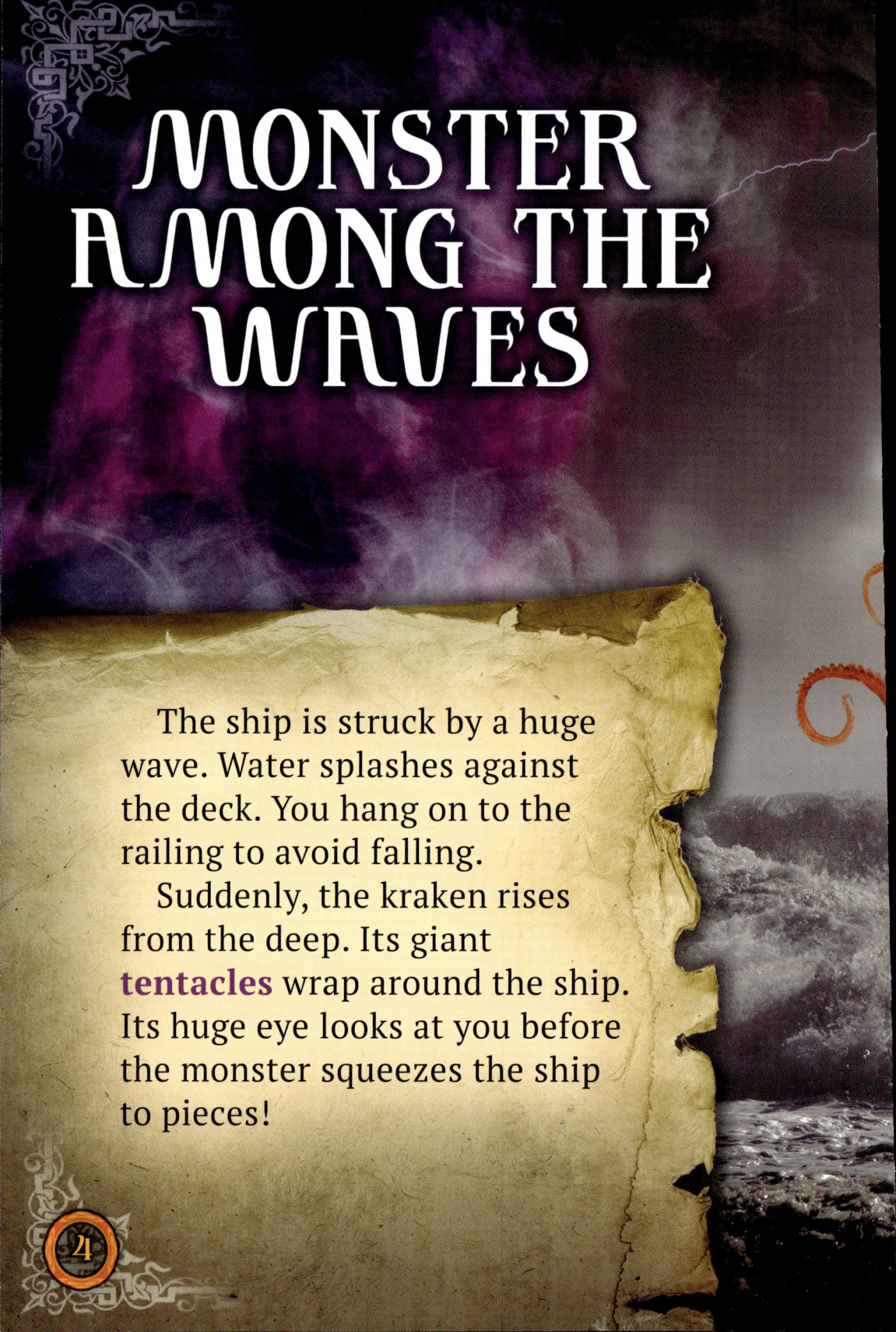

The ship is struck by a huge wave. Water splashes against the deck. You hang on to the railing to avoid falling.

Suddenly, the kraken rises from the deep. Its giant **tentacles** wrap around the ship. Its huge eye looks at you before the monster squeezes the ship to pieces!

5

Danger in the Deep

The kraken likes to stay in the deepest parts of the sea. It only surfaces when it is disturbed.

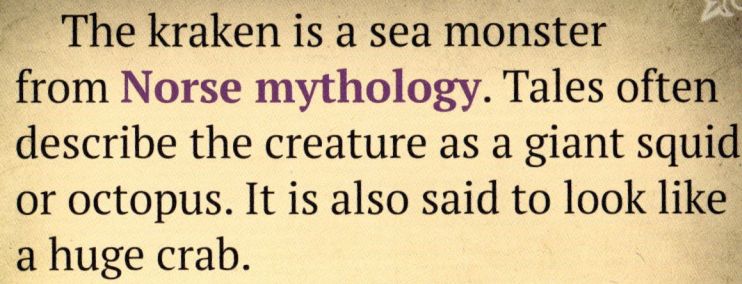

The kraken is a sea monster from **Norse mythology**. Tales often describe the creature as a giant squid or octopus. It is also said to look like a huge crab.

The monster was often mistaken for an island. Old stories mention sailors who tried to land on it. But they discovered it was alive!

Kraken Origin

Iceland
Sweden
Norway
Denmark
Europe

Norse countries =

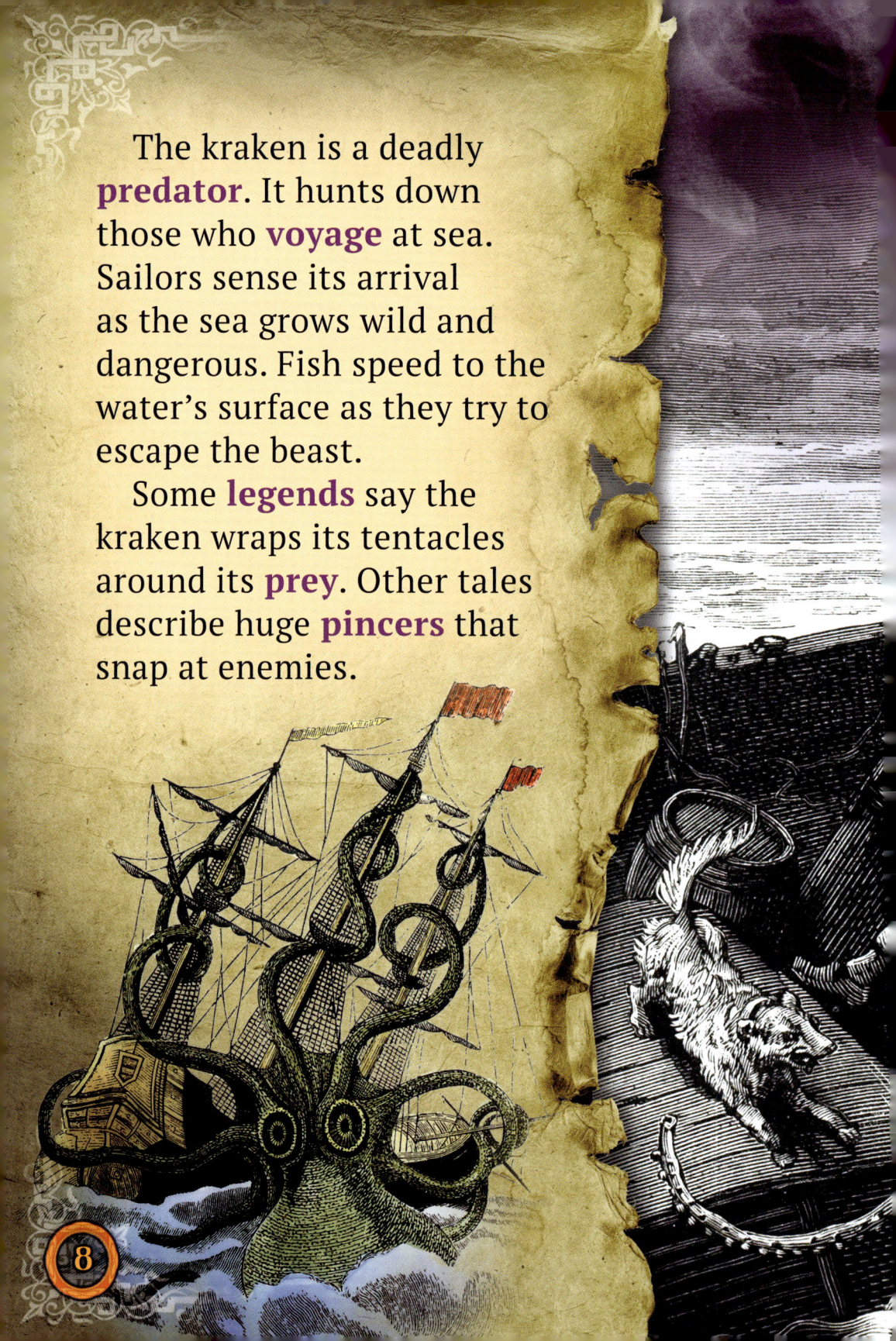

The kraken is a deadly **predator**. It hunts down those who **voyage** at sea. Sailors sense its arrival as the sea grows wild and dangerous. Fish speed to the water's surface as they try to escape the beast.

Some **legends** say the kraken wraps its tentacles around its **prey**. Other tales describe huge **pincers** that snap at enemies.

9

TALES AND SIGHTINGS

The myth of the kraken likely began in 1180. King Sverre of Norway wrote about the sea's many monsters. He described the kraken as one of the largest.

King Sverre of Norway

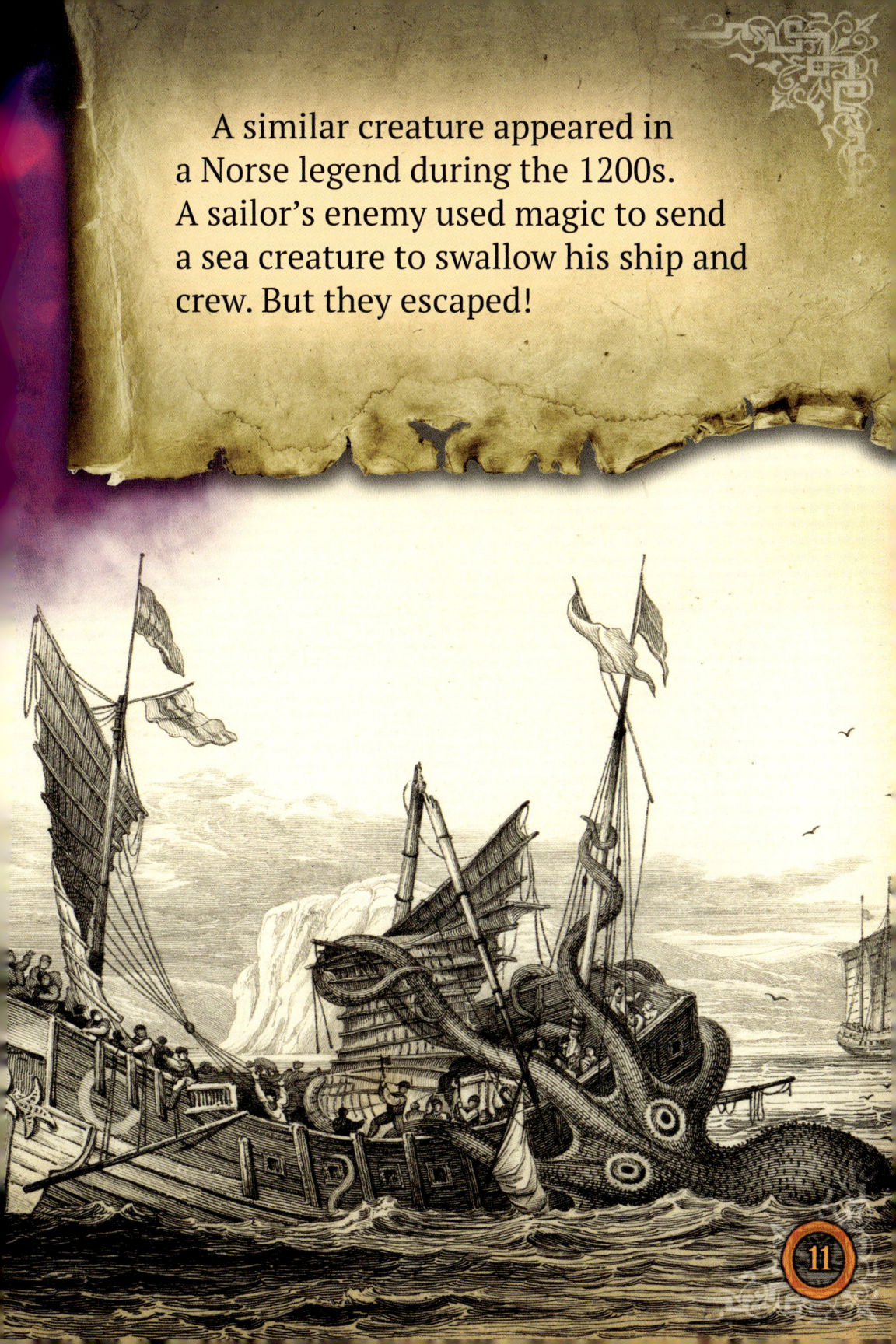

A similar creature appeared in a Norse legend during the 1200s. A sailor's enemy used magic to send a sea creature to swallow his ship and crew. But they escaped!

More kraken tales popped up as sea voyages became more common. Olaus Magnus created a map of the sea between 1527 and 1539. It included pictures of creatures found in the waters.

Magnus wrote about the creatures on his map in 1555. He said the kraken was a fish with long horns. It had red eyes. It had hairs like feathers and a long beard.

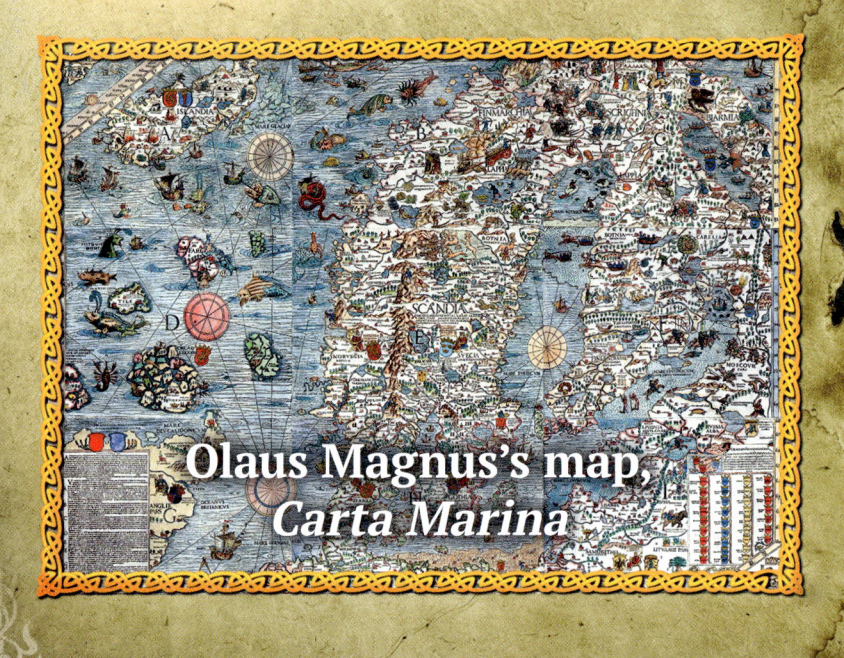

Olaus Magnus's map, *Carta Marina*

drawings of sea monsters described by Olaus Magnus

Similar Creatures

Hydra    bishop-fish    Leviathan

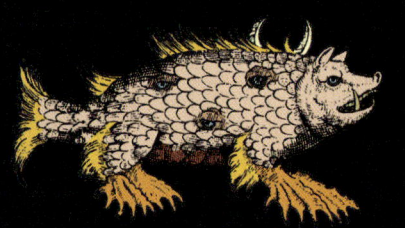

sea pig    sea devil

The kraken was described as the largest animal in the world in a 1752 tale. It was thought to be as big as mountains or islands.

The tale tells of a crew that landed on an uncharted island. They were shocked when the ground moved beneath their feet. The sailors had no idea that they had set foot on the kraken!

Swirling Waters

Many people believed the kraken created large whirlpools. These waters could pull a ship to the bottom of the sea!

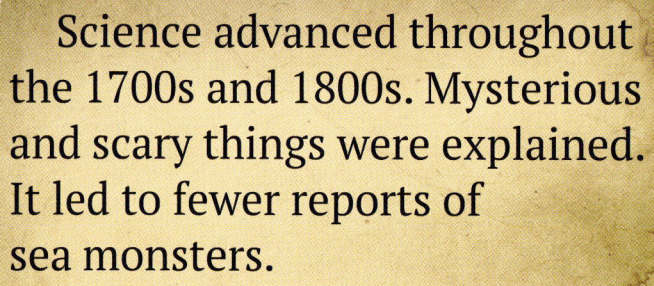

Science advanced throughout the 1700s and 1800s. Mysterious and scary things were explained. It led to fewer reports of sea monsters.

In 1857, Japetus Steenstrup studied a giant squid beak that washed ashore in Denmark. He concluded that sightings of the kraken were actually a giant squid!

Squid Stats

The largest giant squid measured 59 feet (18 meters) in length.

giant squid beak

Kraken Timeline

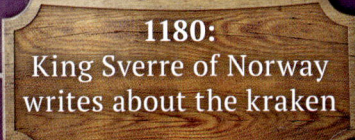

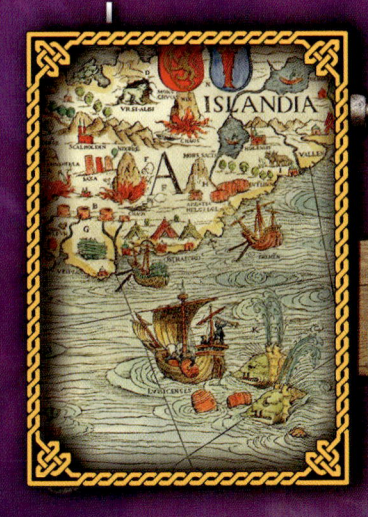

1180: King Sverre of Norway writes about the kraken

1527 to 1539: Olaus Magnus creates a sea monster map called *Carta Marina*

1857: Japetus Steenstrup concludes the giant squid may be the mythical kraken

THE KRAKEN LIVES ON

The kraken is still around today! SeaWorld Orlando is home to the Kraken roller coaster. Riders speed along tracks that reach more than 150 feet (46 meters) above ground!

In 2020, Seattle named its new hockey team the Kraken. Many once believed the creature lived in the Pacific Ocean around the city!

Seattle Kraken flag

**SeaWorld Orlando
Kraken roller coaster**

The creature also appears in video games, books, and movies. The video game *Splatoon* features a tool called Kraken. It turns the player into a giant squid. The kraken also appears in the movie *Pirates of the Caribbean: Dead Man's Chest*.

The kraken is still remembered as a fearsome mythical creature. The monster keeps pulling us in!

Pirates of the Caribbean: Dead Man's Chest

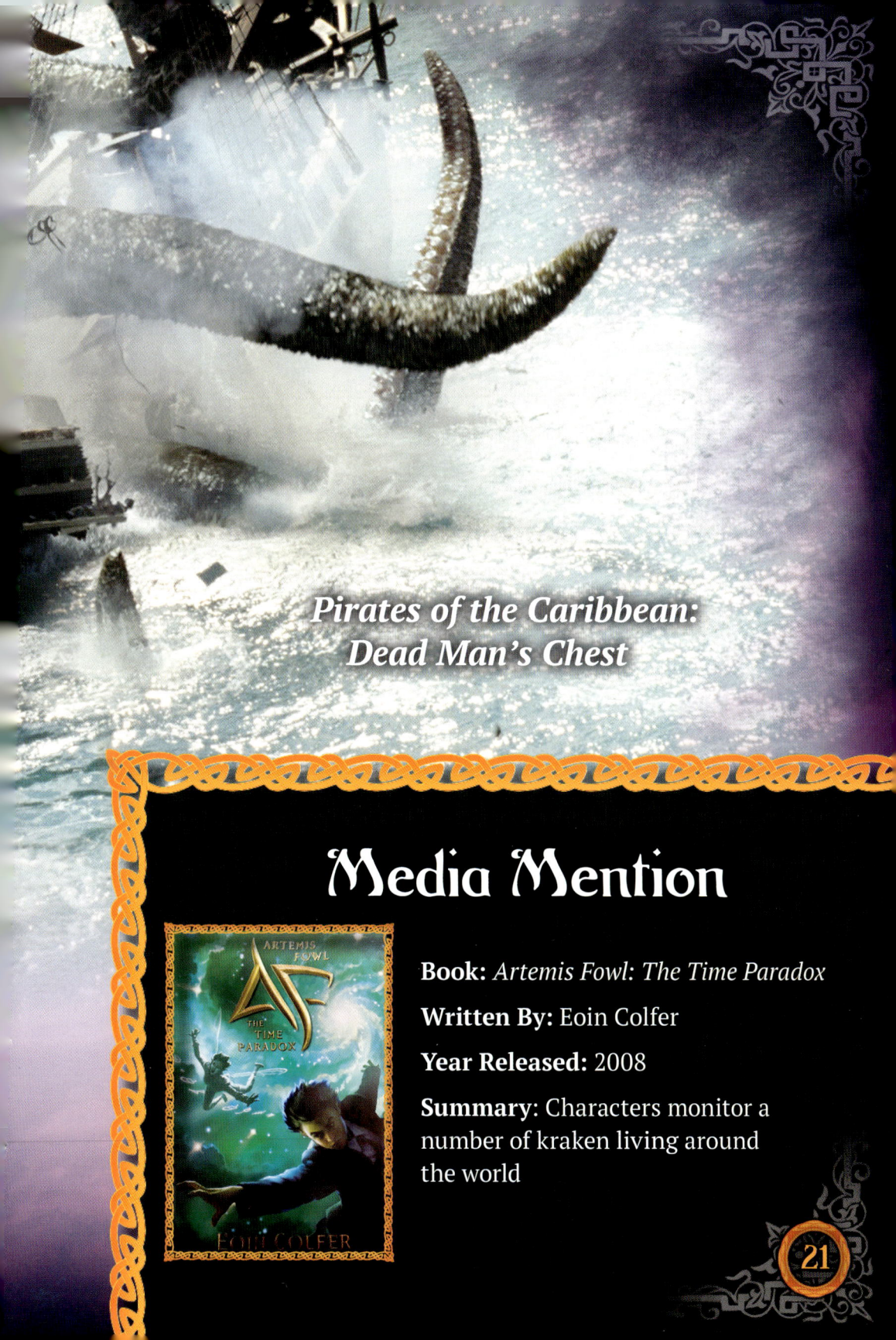

Media Mention

Book: *Artemis Fowl: The Time Paradox*

Written By: Eoin Colfer

Year Released: 2008

Summary: Characters monitor a number of kraken living around the world

GLOSSARY

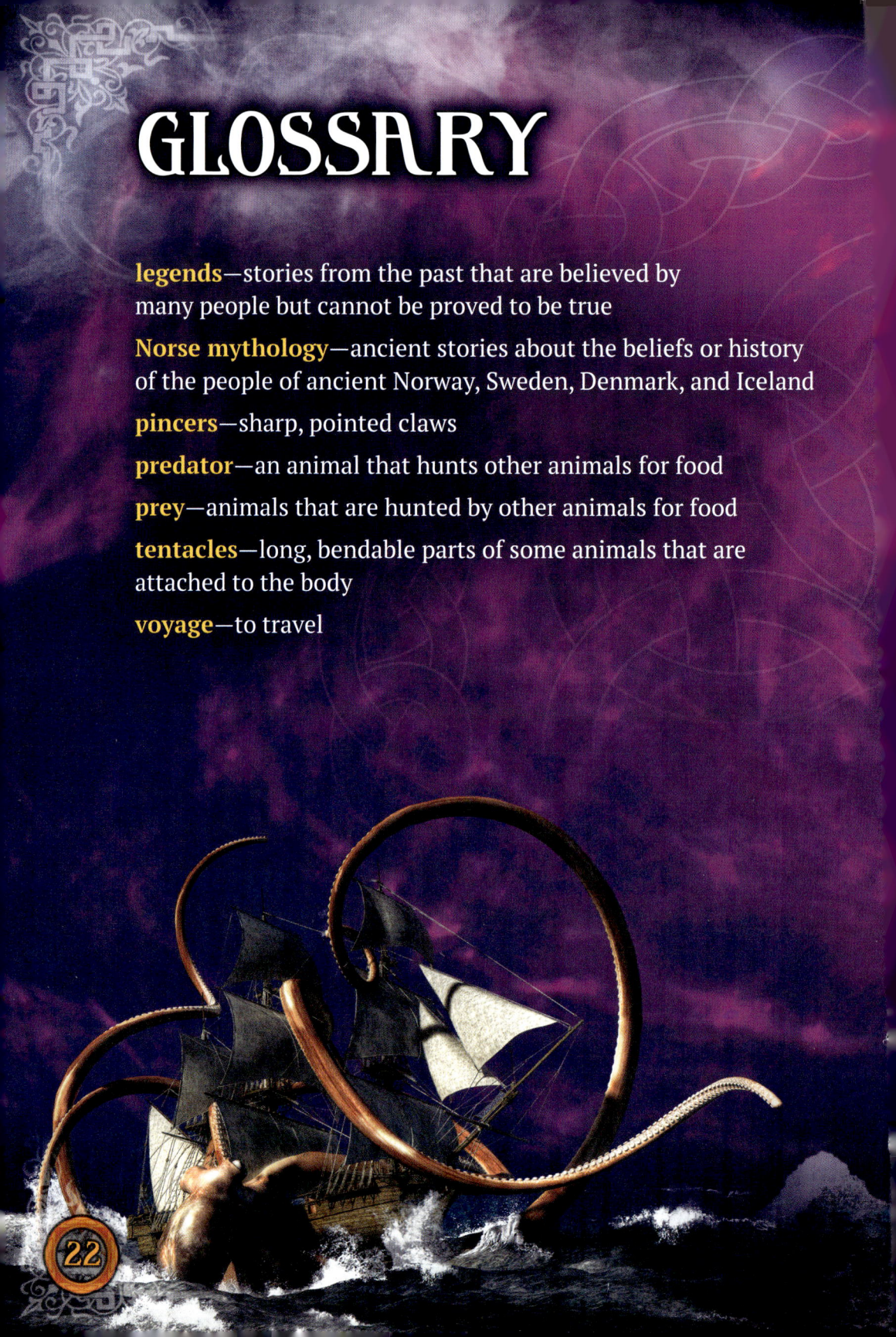

legends—stories from the past that are believed by many people but cannot be proved to be true

Norse mythology—ancient stories about the beliefs or history of the people of ancient Norway, Sweden, Denmark, and Iceland

pincers—sharp, pointed claws

predator—an animal that hunts other animals for food

prey—animals that are hunted by other animals for food

tentacles—long, bendable parts of some animals that are attached to the body

voyage—to travel

TO LEARN MORE

AT THE LIBRARY

Goddu, Krystyna Poray. *Sea Monsters: From Kraken to Nessie*. Minneapolis, Minn.: Lerner Publications, 2017.

Lawrence, Sandra, and Stuart Hill. *The Atlas of Monsters: Mythical Creatures from Around the World*. Philadelphia, Pa.: Running Press Kids, 2019.

Sautter, A.J. *Discover Harpies, Minotaurs, and Other Mythical Fantasy Beasts*. North Mankato, Minn.: Capstone Press, 2018.

ON THE WEB

FACTSURFER

Factsurfer.com gives you a safe, fun way to find more information.

1. Go to www.factsurfer.com

2. Enter "kraken" into the search box and click 🔍.

3. Select your book cover to see a list of related content.

INDEX

appearance, 7, 8, 12

Artemis Fowl: The Time Paradox, 21

Denmark, 7, 16

explanations, 16

giant squid, 7, 16, 20

history, 7, 10, 11, 12, 14, 15, 16, 18

King Sverre, 10

Kraken roller coaster, 18, 19

legends, 8, 11

Magnus, Olaus, 12, 13

mythology, 7, 10, 20

Norway, 7, 10

origin, 7

pincers, 8

Pirates of the Caribbean: Dead Man's Chest, 20, 21

predator, 8

sailors, 7, 8, 11, 15

sea, 6, 8, 10, 12, 15

Seattle Kraken, 18

similar creatures, 14

size, 10, 14, 16

Splatoon, 20

Steenstrup, Japetus, 16

tentacles, 4, 8

timeline, 16-17

whirlpools, 15

The images in this book are reproduced through the courtesy of: sko1970, front cover (tentacles); Kalifer - Art Creations, front cover (hero); A.Dina, p. 3; andrey polivanov, p. 3 (background); muratart, pp. 4-5 (background); zhengzaishura, pp. 4-5 (tentacles); Science History Images/ Alamy, pp. 6-7, 14 (bishop-fish, sea devil, sea pig); Universal History Archive/ Getty, pp. 8-9; North Wind Picture Archives/ Alamy, p. 9; Chronicle/ Alamy, pp. 10, 17 (top); Album/ Alamy, pp. 10-11; Olaus Magnus/ Wiki Commons, pp. 12, 17 (middle); Bettmann/ Getty, pp. 12-13; Interfoto/ Alamy, p. 14 (Hydra); The Picture Art Collection/ Alamy, p. 14 (Leviathan); duncan 1890/ Getty, pp. 14-15; FLPA, p. 16; August Jerndorff/ Wiki Commons, p. 17 (bottom); 400tmax, p. 18; viaval, pp. 18-19; Moviestore Collection Ltd/ Alamy, pp. 20-21; Bellwether Media, p. 21; Daniel Eskridge, p. 22.

Prospect Heights Public Library
12 N. Elm Street
Prospect Heights, IL 60070
www.phpl.info